Chemical Engineering & the Law Forum 2024 and Liaison Functions 2024

Held at the 2024 AIChE Spring Meeting and 20th Global Congress on Process Safety

New Orleans, Louisiana, USA
24-28 March 2024

ISBN: 979-8-3313-1707-2

Printed from e-media with permission by:

Curran Associates, Inc.
57 Morehouse Lane
Red Hook, NY 12571

Some format issues inherent in the e-media version may also appear in this print version.

Printed with permission by Curran Associates, Inc. (2025)

For permission requests, please contact AIChE
at the address below.

AIChE
120 Wall Street, FL 23
New York, NY 10005-4020

Phone: (800) 242-4363
Fax: (203) 775-5177

www.aiche.org

Additional copies of this publication are available from:

Curran Associates, Inc.
57 Morehouse Lane
Red Hook, NY 12571 USA
Phone: 845-758-0400
Fax: 845-758-2633
Email: curran@proceedings.com
Web: www.proceedings.com

TABLE OF CONTENTS

61a Chemical Processes and Trade Secret Protection

Charles Collins-Chase, Paul Townsend

Chemical processes are the backbone to numerous products making their way into the market in many industries. Despite the large chemical processes industry, intellectual property protection for such processes is less than clear. Companies may use patents to protect certain chemical processes; however, patent protection requires public disclosure of the process. If a company does not wish to publicly disclose its chemical process, it is left without the availability of patent protection. How then do companies keep chemical processes protected? This presentation will provide an overview of the basics of IP, specifics on trade secret law, and important trade secrets considerations for companies wishing to protect chemical processes. Speakers will also highlight several case studies addressing chemical processes trade secrets and lessons learned from those cases.

61b Patent Litigation Trends in the US and Beyond

Kaylie Everett, Steven Chao

This presentation provides an overview of patent litigation trends within the US and globally. Patent litigation is a topic of focus for the legal departments of many companies. Non-practicing entities, or NPEs, account for a large percentage of patent litigation each year. This presentation will explore trends in patent litigation filings by both operating companies and NPEs within the US. Many companies do not only have domestic operations and their patent litigation risk extends globally. This presentation will provide insight into the global patent filing and patent litigation landscape.

Machine Learning for Radioactive Source Detection Using Maximum Detection Distance in Environmental Remediation

Nancy Ibrahim, Amin Amirlatifi
Dave C. Swalm School of Chemical Engineering, Mississippi State University, MS 39762

Abstract

The critical task of detecting radioactivity in suspected areas of contamination is essential for safeguarding public safety and minimizing potential adverse consequences. This study centers around the adaptation of a Maximum Detection Distance (MDD) algorithm, harmonized with the localization function of a mobile gamma-ray detection system. When exploring specific routes, such as former roadways and earthen berms, the ability to detect a radioactive source is inherently tied to the distance from the path. Therefore, understanding the MDD is critical for accurately assessing and declaring an area "clean".

The maximum detectable distance signifies the farthest distance achievable between the vehicle and the radiation source under specific source characteristics and activity levels. The results demonstrate the algorithm's favorable accuracy in predicting MDD for a 2"x2" NaI(Tl) detector, with optimal acquisition time intervals of 5 to 10 seconds for Cs-137 unshielded point sources. Notably, longer acquisition time intervals exceeding 20 seconds prove more effective for detecting higher activity sources, whereas shorter intervals enhance MDD at the same speed. However, to increase the MDD, shorter time intervals should be selected when operating at higher speeds. These findings provide valuable guidance for optimizing the detector settings and vehicle speed in mobile searching scenarios. Ultimately, the outcome of this study holds the potential to bridge the chasm between theoretical advancements and their practical implementation. Emergency remedial clean-up teams can utilize these findings to optimize their vehicle routes, ensuring comprehensive coverage of environmental monitoring areas where concerns of potential contamination may arise.

Keywords: Machine Learning, Gamma-ray detection, Mobile detection

1. Introduction

1.1 Motivation

Detection of radioactivity in suspected areas of contamination plays a crucial role in safeguarding public safety and mitigating potential adverse impact. When a particular area is suspected to be contaminated with one or more sources of radioactivity, the ability to identify their precise location(s) and magnitude is crucial. The extent of detecting various radiation sources is well established for both detection devices and detection methodology. Nevertheless, pinpointing and locating low-level radiation sources present significant challenges, especially when compared to high-intensity sources. These challenges stem from two primary factors: fluctuations in background radiation levels and the probabilistic nature of radiation measurements.

The value of detecting radioactivity becomes more pronounced in scenarios involving potential threats, such as dirty bombs. The prompt and precise detection of these sources is vital not only to avert immediate casualties resulting from an explosion, but also to mitigate the long-term health repercussions of radiation exposure. Identifying the location and intensity of radioactive sources early can help control the spread of contamination.

In the realm of radiation detection, the choice between a laboratory setting and field deployment introduces distinct considerations. When conducting gamma radiation measurements in the field, particularly in emergency scenarios or environmental monitoring, sodium iodide (NaI) detectors are often favoured over high-purity germanium (HPGe) detectors for practical reasons. However, a significant challenge is the high contribution of natural backgrounds to measurements due to the detectors' sensitivity to gamma emitters. This sensitivity makes it tough to detect sample activities below this background.

In field applications, radiation is measured with a mobile detector generally mounted on a vehicle. Examples of mobile gamma detectors are Geiger-Muller counters and sodium iodide (NaI) detectors. Although NaI detectors exhibit lower resolution and sensitivity, their affordability and portability make them more accessible and suitable for rapid response scenarios. Though, the challenges become more pronounced when dealing with low-level sources or situations where multiple sources are involved. The unpredictable nature of background radiation and the probabilistic characteristics of radioactive measurements make it challenging to set fixed thresholds for detection. Moreover, accurately estimating the location and strength of sources becomes intricate, especially in scenarios involving multiple dispersed sensors.

In the realm of detecting radioactive sources, substantial challenges such as limited detection range or sensitivity, environmental variability, and radiological interferences require more effective remediation procedures. In the event of a nuclear fallout or spill, radioactivity can be dispersed across diverse environments, including urban areas, industrial sites, and natural landscapes. Mobile detection systems must navigate through these environments while efficiently detecting and localizing sources amidst background radiation and other environmental factors. Subsequently, achieving sufficient sensitivity to detect low-level radioactive sources can be difficult, particularly in the presence of background radiation due to natural and anthropogenic sources. The objective of this study is to provide valuable insights into the maximum distance at which a radioactive source can be detected, enabling operators to make informed decisions in real-time. By integrating data from mobile gamma-ray detectors and GPS positioning systems, this study can support rapid response, minimizing the time between detection and action to mitigate potential risks to public safety.

The need for reliable and comprehensive remediation procedures is crucial due to the potential risks to the public health and environment. Whether stemming from historical mining activities or industrial waste disposal, the contaminants involved can pose severe health hazards. There have been several abandoned or decommissioned mining sites in New Jersey that have led to remediation procedures due to the environmental hazards they pose. The clean-up efforts for the Ringwood Mines/Landfill site began in the 1980s and continues to the present, as it was used for dumping industrial waste, including paint sludge and other hazardous materials. The clean-up involved the removal and proper disposal of contaminated soil and waste, as well as capping and monitoring the landfill. The Boeing Michigan Aeronautical Research Center (BOMARC) Missile Accident site also located in New Jersey was a location where a significant nuclear accident occurred in 1960. The accident involved a BOMARC missile, which was equipped with a nuclear warhead, and resulted in the release of radioactive materials into the environment. A significant area around the accident site was contaminated, necessitating extensive cleanup efforts. The incident raised concerns about the safety and security of nuclear weapons and their maintenance procedures.

In these cases, and many others, the need for dependable remediation procedures is crucial due to the potential risks to public health and the environment. The scope of the remedial investigative multi-phase field data used in this study entailed radiological characterization surveys to determine the nature and extent of technologically enhanced naturally occurring radioactive materials (TENORM), to support the development of remedial alternatives. The main purpose of this study is to extend the adaptation of a Maximum Detection Distances (MDD) calculation algorithm, harmonized with the localization function of a mobile gamma-ray detection system. When exploring specific routes, such as former roadways and earthen berms, the ability to detect a radioactive source is constrained by the distance from the path – therefore, knowing the MDD is significant to declare an area "clean".

2. Related Work

Over the past several decades, researchers have primarily concentrated their efforts on the challenge of pinpointing the locations of radioactive sources through data collected by radiological detectors within the realms of nuclear physics and radiological monitoring. Much of the research to date has been centred on the localization of single radioactive sources within various planes using detectors.

In this study, we delve into the estimation of parameters related to radioactive sources under these specific conditions. In terms of location accuracy, the average error in estimating location parameters in our study does not significantly deviate from the findings of other studies focusing on single-source localization.

While many studies have delved into the art of radiation source localization, a considerable portion of the extant literature revolves around methods that excel in locating radiation sources but fall short of effectively distinguishing between various source types. A multitude of methodologies, from [8] investigations into single-source localization through least square methods or maximum likelihood estimation to [13] adaptation of the Time Difference-of-Arrival (TDOA) algorithm for identifying low-level point radiation sources, focus predominantly on pinpointing sources. There have been efforts, such as [8] work employing mobile robots for radiation source detection and localization, aimed at maximizing information gain during search operations.

[12] applied maximum likelihood and Bayesian methodologies for locating multiple radioactive sources, particularly in a 2-D plane. [17] applied the Bayesian principle and sequential Monte Carlo methods to estimate radiation source parameters --which serves the same purpose in our precept-- showcasing similar localization accuracy to other single-source location studies.

Traditionally, the efficiency of radiation detectors is assessed in controlled environments with uniform radiation distribution and consistent setups. However, radiation fields encountered in mapping scenarios often exhibit variations in intensity and spatial distribution. Research has indicated that the placement of sampling points can significantly influence the accuracy and effectiveness of radiation detection systems [14]. Insufficient sampling points may result in missed detection of radioactivity, thus posing risks of radiation hazards and exposures.

Previous studies have addressed this issue by fixing the distance between sampling points based on expected or known source intensities [2][6]. Yet still in scenarios involving unknown radiation fields, a strategy is necessary to determine the positioning of sampling points and assess their sensitivity to radiation levels.

In the context of this research study, mobile gamma spectrometry entails the deployment of one or multiple gamma spectrometers on a mobile platform, often in the form of a vehicle, such as an automobile. A fundamental challenge associated with mobile gamma spectrometry lies in the detection of faint radiation signals originating from distant sources amidst the prevalent natural background radiation. This challenge can be attributed, in part, to the relatively brief data acquisition intervals, which are influenced by the speed at which the vehicle is traveling. [8] have effectively employed statistical analysis of collected data to discern faint signals originating from lost radiation sources while minimizing false-positive readings.

In a prior study by [14] the minimum detectable amount (MDA) was proposed as a benchmark to gauge the effectiveness of radiation monitoring algorithms – defining the detection limit as the coverage and sensitivity of the sampling point. MDA represents the smallest activity or intensity level detectable by the instrument that is not contributed from background radiation (7). Decreasing MDA values signify heightened measurement sensitivity across similar instruments [15]. In an earlier study, Song et al. computed the MDA for their system comprising scintillation detectors positioned at each vertex of a rectangular "Region of Interest" (ROI) [16]. This computation factored in the geometry of the ROI and the detector positions, indicating the minimum quantity of radioisotope detectable by the system.

Further enhancements to search methodologies have been proposed by [4] who applied Bayesian statistical techniques to determine both the distance and activity of detected gamma-ray sources as the measuring vehicle passes. These research endeavours have primarily concentrated on localizing orphan gamma-ray sources within extensive areas, prompting inquiries into the maximum detection distances (MDD). The primary focus on [4] study was on developing an algorithm to calculate MDD for mobile gamma-ray detection systems in the context of locating lost radiation sources. [4] highlighted as the detector count may exceed the alarm level at different points along its path can make calculating MDD complex when the detector is in motion [10]. This concept was adapted as the localization function of this study.

A distinguishing factor of this study lies in its reliance on real-world data collected, in contrast to the prevalent use of simulation-based data in much existing studies. This reliance on empirical data collected from the field not only aligns the research more closely with practical applications, but also allows for a more accurate representation of the complexities and nuances present in real radiological monitoring scenarios. This study's approach allows one to pinpoint critical factors affecting detection accuracy and tailor optimization strategies accordingly. Consequently, the outcomes of this study hold the potential to bridge the gap between theoretical developments and practical implementation, making radiological monitoring technology more accessible and cost-effective for a broader range of applications.

3. Methodology

3.1 Detection Classification

The initial phase of analysing a potentially radioactive region involves conducting a comprehensive search to pinpoint the locations of any existing radioactive hotspots. To perform an inspection in such radioactive scenarios, portable devices, such as radiological sensors are utilized. These sensors can be hand-held, integrated into a backpack, or even embedded in mobile robots like unmanned aerial vehicles (UAVs) and ATVs.

The detection approach that will be employed in this model is centered around binary classification to distinguish between detected radioactive sources and non-detected scenarios. The database used for training and evaluation is constructed from real-life survey points collected by a 4-wheel drive ATV equipped with GPS and a sodium iodide activated with thallium (NaI [Tl]) gamma scintillation detector.

The decision to use a NaI(Tl) gamma scintillation detector in field radiation detection is influenced by several factors, primarily its well-established performance characteristics and suitability for various applications. NaI(Tl) detectors offer high sensitivity and energy resolution, making them effective for detecting and identifying gamma-ray sources across a broad energy range. Their relatively low cost compared to other types of detectors, such as HPGe detectors, also makes them accessible for many field monitoring applications. One of the key advantages of NaI(Tl) detectors is their versatility and portability, allowing for easy deployment in diverse environments, including field surveys, border security, and emergency response scenarios.

The model will utilize a series of calculations and statistical methodologies to assess the presence of radioactive sources within a given scenario. Through this process, the model will also compute an average background count rate, establishing a fundamental benchmark for radiation levels within the given area. By applying statistical techniques, such as the determination of critical level (C_L), detection limit (L_D), and MDA, the model can accurately delineate the threshold of detectability with a desired 95% level of confidence.

The critical level and detection limits serve as fundamental parameters in assessing the sensitivity and reliability of radiation detection systems. The critical level represents the threshold net count in a zero-mean count distribution that corresponds to a predefined probability of being exceeded. This parameter is pivotal in distinguishing between background radiation fluctuations and signals from radioactive sources. By statistically defining the critical level, we can establish a reliable criterion for detecting anomalous radiation events amidst varying levels of environmental background radiation. The equation used in this study for C_L is defined as [9]:

$$C_L = 1.645\sqrt{2S_b^2} = 2.33S_B \qquad (1)$$

The L_D is computed as the number of mean net counts obtained from samples that are highly likely to surpass the C_L. In essence, the detection limit denotes the minimum radiation signal strength that can be reliably detected above the background noise. Samples with net counts exceeding the detection limit are deemed to possess significant radiation signals warranting further investigation, while those falling below the limit are likely to represent low-level sources or background noise. The L_D is given below in equation 2.

$$L_D = 3 + 4.65S_B \qquad (2)$$

The MDA of a sample follows directly from the detection limit concepts. The MDA of a radionuclide depends on several factors; and for nonequal background and sample counting times, the MDA at a particular gamma-ray energy is defined as [9]:

$$MDA = \frac{3+3.29[B_r \cdot t_s\left(1+\frac{t_s}{t_b}\right)]^{\frac{1}{2}}}{\varepsilon \cdot S \cdot Y \cdot t_s} \quad (3)$$

where,

B_r = the gamma-ray count rate for an uncontaminated sample (in counts per second), this is also referred to as the background rate,

t_s = contaminated soil sample counting time in seconds,

t_b = uncontaminated soil sample counting time in seconds,

ε = detector efficiency in counts per emitted gamma-ray,

S = contaminated soil sample mass in kilograms; and

Y = number of gamma-ray emissions per radionuclide disintegration

Therefore, the smallest MDA achievable by the given detection system depends on the background rate, efficiency based on sample geometry, total mass, and the contaminated and uncontaminated sample counting time(s).

By filtering out data points where the net counts exceed the L_c, the focus is directed towards identifying low-level radioactive sources amidst typical background radiation levels. This filtering process ensures that subsequent calculations prioritize the detection of sources with elevated activity levels while minimizing the inclusion of potential hot spots. The detection status of the measured data is then determined based on whether the measured activity surpasses the critical level, as shown as:

1 (detected), if measured activity $\geq L_c$
0 (not detected), if measured activity $\leq L_c$

These statistical methods, grounded in the calculation of critical levels and detection limits, are integral to the accurate identification of radioactive sources and are applied systematically within the model to optimize detection performance.

3.2 Theory and Application

In the realm of in situ gamma spectrometry, [1] laid out the groundwork for stationary detectors. However, when the spectrometer is in motion during measurements, accounting for the fluctuating source-detector distance becomes imperative. For mobile gamma spectrometry, where both the radiation source and the detector are in motion, Lloyd A. Currie's method for a priori calculating the detection limit in stationary scenarios lacks direct applicability to mobile gamma spectrometry.

The calculation for MDD, introduced by [4], combines the principles of radiation transport, interaction, and statistical laws to determine the maximum detection distances achievable for photon point sources during mobile gamma spectrometry. These calculations address scenarios involving both stationary measurements and those taken while in motion along a linear trajectory past a radiation source.

The MDD value denotes the maximum distance from a source at which the detection system, comprising gamma scintillation detectors, can effectively identify and classify the presence of that source. For stationary detectors, the calculation of MDD (R_{max}=R_{fix}) to a point source is straightforward [9]. The primary photon fluence rate, ϕ, in air at distance r from a source can be expressed using Equation 4, where the linear attenuation coefficient for photon absorption in air is represented as μ_a.

$$\phi = \frac{\dot{S}e^{-\mu_a r}}{4\pi r^2} \quad (4)$$

The inverse square law ($4\pi r^2$) signifies the reduction in photon fluence as a person distance themselves from the radiation source. The exponential factor ($e^{-\mu}$) becomes more crucial at greater distances, emphasizing the impact of air attenuation.

Alternatively, for a moving detector, the distance, *r*, changes with time. As it is moving in a straight direction with constant speed, the distance between the detector and the source varies in time passing a point source located at a certain distance, *R*, perpendicular to the path shown in Equation 5. The constant speed of the detector and vehicle is represented as *v* and *t* is the time.

$$r(t) = \sqrt{R^2 + (vt)^2} \quad (5)$$

The time variation of the primary photon fluence rate at the detector in mobile measurements occurs when Equation 4 is inserted into Equation 5, yielding Equation 6.

$$\varphi(t) = \frac{Se^{-\mu\sqrt{R^2+(vt)^2}}}{4\pi(R^2+(vt)^2)} \quad (6)$$

Since a gamma spectrometer's detector doesn't capture all incoming photons, the detector's efficiency, denoted as ε_{90}, is a key factor in detecting low levels of radiation from distant sources. Equation 7 represents the effective area of an imaginary detector that absorbs all incident photons. This is measured by placing known radioactive sources some distance away and noting the full energy peak count rates.

$$\epsilon_{90} = \frac{N_{90}}{\varphi} \quad (7)$$

where,
N_{90} = full energy peak count rate from photons incident at 90° to the right of the driving direction
φ = primary photon fluence rate at the detector from a source

As a detector-equipped vehicle moves along a road past a radiation source, it receives primary photons from various angles, and therefore, the detector's effectiveness can change based on the angle of photon incidence. The relative angular efficiency of a detector, denoted as δ_θ, and shown in Equation 8, quantifies this variation comparing the count rate at a specific angle θ to that at 90°.

$$\delta_\theta = \frac{N_\theta}{N_{90}} \quad (8)$$

where,
N_θ = count rate at angle θ varying from 0 to 180°

To relate this relative angular efficiency variation to practical measurements, Equation 9 is used to connect the primary photon fluence rate (φ_θ) at a specific angle to the net count in the gamma spectrometer. Taking into consideration factors like effective detector area ($\mathcal{E}_\theta$) and the acquisition time interval, this formula accounts for the detector's effectiveness changes with the angle of incoming photons.

$$\varphi_\theta = \frac{N_\theta}{\varepsilon_\theta \Delta t} \quad (9)$$

The traditional approach to measure the activity of a radionuclide in gamma spectroscopy is to count the pulses registered in a full energy peak in the pulse height distribution. In the realm of mobile gamma spectroscopy, due to a generally short the acquisition time, the full energy peak area can be consequently low where a distinct peak isn't clearly present. A more straightforward approach, using the ROI method, is more applicable as it requires minimal time and effort whether a distinct peak is present.

The net contribution from a source can be obtained by adding the number of pulses within a specific photon energy interval and subtracting the average background counts in the ROI. The probability of obtaining X pulse counts in the ROI follows the Poisson distribution, where each probability represents the likelihood of observing the specified number of counts in the ROI during the acquisition time interval [9]. A simplified form of the distribution is shown in Equation 10.

$$P(x) = \frac{(\overline{x})^x e^{-\overline{x}}}{x!} \quad (10)$$

3.3 Real Data

The fundamental unit for measuring radioactivity is becquerel (*Bq*), which signifies the number of nuclear disintegrations occurring in one second. When it comes to measuring intensity -the number of particles detected during a specific time period- in each area, the unit used is Bq/m^2 [9]. However, in real world applications, unlike in controlled simulations, actual sensors may not always provide measurements in the most convenient unit. They simply count the interactions of detected particles without assessing the energy of these particles. In the case of a NaI(Tl) scintillator, this device identifies ionizing particles using a NaI crystal connected to a photomultiplier tube and counts the number of these particles within a defined time frame. Therefore, since a NaI(Tl) scintillator has no way to measure the energy of the particles, the measurements are given in units of counts per second (CPS) or counts per minute (CPM).

A radiological field survey was carried out for the Heritage Minerals, Inc. (HMI) site located in Manchester Township, New Jersey, covering a total land area approximately 260 acres. A portion of the HMI property was previously used for mineral production, with by-product wastes suspected to constitute TENORMs. The remedial investigation work had been reviewed and approved by the New Jersey Department of Environmental Protection (NJDEP).

The GPS detection system and survey instrument was mounted on an ATV approximately twelve inches above the ground surface. This apparatus setup is depicted in Figure 1. Each were connected to GPS units equipped with data loggers. The GPS system displayed the locations of the data points, which allowed for real-time monitoring of the survey coverage. The gamma survey measurements were made either from an ATV or on foot, depending on the accessibility of the survey areas. An RS-232 output PROM card in the scaler/ratemeter sent the gamma count rates to the GPS data logger at 1-second intervals. The ATV proceeded at a speed of approximately one meter per second, which typically produced about 3,000 to 5,000 gamma measurements per acre, depending upon the accessibility of the terrain. A total of 51,957 samples were collected.

Figure 1. *A Ludlum 44-10 2" x 2" NaI(Tl) detector, suspended from an all-terrain vehicle (ATV) positioned 12 inches above the ground. This setup enables radiation monitoring across varied terrains, at a speed producing 3,000 to 5,000 gamma measurements per acre for this study.*

Based on information provided by the instrument manufacturer, the sensitivity of the Model 44-10 gamma probes used within this investigation is about 900 cpm/μR/hr, based on a Cesium-137 (Cs-137) gamma source. To account for the survey measurements, a precision test was performed with a standard of a known dose rate of 250 uR/hr. The count rate was obtained three times from this standard and then averaged. From the measurement of the known, a conversation rate from cpm to uR/h was calculated and later applied into the machine learning algorithm.

Measuring background radiation levels in undisturbed areas serves as a critical step in establishing baseline values against which the radiation levels in radiologically impacted land can be compared. The background reference areas selected consisted of non-impacted areas that were undisturbed by historic mining or milling operations (and confirmed from historic aerial photographs). Areas with developed plant life, forested or old growth areas, where overgrowth and size of the vegetation and trees indicated this area was well established before the site operation, also indicate that the area remained undisturbed during the operating life of the facility. A total of --51,957 data points were scanned for a true background of undisturbed area on the sire (without excavation), yielding an average reference value of 4737.7 CPM.

3.4 Machine Learning

In this study, we leverage an approach modified by the work of [4] to compute the MDD in mobile gamma-ray detection scenarios. The photon fluence rate, at a constant speed, can be expressed as a function of the distance $d = vt$ along the road, where $d = 0$ at $t = 0$, commonly referred to as the intensity curve. However, since the primary photon fluence rate cannot be used directly for a mobile detector, an iterative numerical approach becomes necessary.

For the calculation, several parameters must be set according to the following:

1. It is assumed that the detector is moving with constant speed along a straight line passing the source at some distance
2. The source is assumed to be unshielded and acting as an isotropic-like point source located on the ground
3. The photon energy for the radionuclide to search for is used to define a region of interest (ROI) in the gamma spectrum
4. The background count rate in the ROI must be determined in advance [9]
5. Source activity is assumed to be stable, so duration of any measurement is negligible when compared to half-life of source

For radioactive sources emitting low levels of radiation, the maximum distance at which an unshielded gamma detector can detect them varies considerably, typically ranging from a few inches up to 15-20 inches under optimal conditions. However, this range is subject to fluctuation due to several factors:

1. Activity of the radioactive source: sources with lower activity emit fewer gamma rays per unit time, thereby diminishing the intensity of radiation at a given distance.

2. Energy of the gamma rays: gamma-rays with lower energy levels are more susceptible to attenuation by air and other materials, resulting in a reduced detection range compared to higher energy gamma rays.
3. Background radiation: elevated levels of background radiation can obscure the gamma radiation emitted by the source, decreasing the effective detection distance.
4. Sensitivity of the detector: different detectors exhibit varying levels of sensitivity, with some capable of detecting gamma radiation over greater distances than others.
5. Environmental conditions: elements such as air humidity, temperature, and atmospheric composition can impact the attenuation of gamma rays, consequently affecting the detection range.

Given the influence of these factors, it is crucial to conduct empirical tests or simulations specific to one's detector and source configuration to accurately establish the maximum detection distance. This empirical approach provides precise insights into the capabilities of the setup under different operating conditions.

The process employed to compute the MDD, denoted as R_{fix}, for an immobile detector within a single acquisition time interval, Δt, involves iteratively refining the distance parameter to reach the desired detection probability threshold, $P_{desired}$. The steps for calculating the maximum detection distances for an immobile detector are as follows:

Input:

- the effective detector area, $\mathcal{E}_\Theta$, as a function of angle $0 < \Theta < 180°$
- the acquisition time interval, Δt
- the average background count rate, $\dot{B}$, in the ROI
- the alarm level, a (counts), relevant for the time interval, Δt, and ROI
- the desired detection probability, $P(X \geq a)|$ background and source
- the activity A of the source, and the branching ratio n_γ

Calculate:

- the average background count, $B = \dot{B}\,\Delta t$
- the mean count, λ, of the Poisson distribution of counts by fitting the previous equation into the integral 1 - P (X $\geq$ a) using the iterative method explained below
- the mean count from the source, $N = \lambda - B$
- the primary fluence rate, ϕ_θ, at the detector
- the probability of false positive detection in the absence of a source from the Poisson distribution (by setting λ = B) and calculating the integral (1– (1 - P (X $\geq$ a))| background

MDD Steps:

- define the initial range for R_{fix}
- execute an iterative loop to converge on the value R_{fix}, in each iteration
- calculate the midpoint of R_{fix} between R_{min} and R_{max}
- calculate the linear attenuation coefficient μ_a using the given function: $\mu_a = (\mu/p)_a p_a$ where p_a is air density and $(\mu/p)_a$ is the mass attenuation coefficient
- calculate the primary photon fluence rate ϕ at R_{fix}
- estimate the detection probability at R_{fix} based on the primary fluence rate and acquisition time
- if the difference between R_{max} and R_{min} is less than the tolerance value, break out of the loop
- if the detection probability exceeds the desired threshold $P_{desired}$, update R_{max} to R_{fix} ; otherwise, update R_{min} to R_{fix}

When employing a moving detector, the distance r varies over time, while the photon fluence rate adheres to an intensity curve outlined by Equation 6. This complicates the calculation of the maximum detection distance, R_{max}, since direct application of Equation 4 is not possible. Instead, an iterative numerical approach must be employed – the iterative calculations are done according to the following steps:

Input:

- all input parameters for immobile detector
- the speed of the detector/vehicle, v

Calculate:

1. the MDD, $R_{fix} = r(0)$, for immobile detector and set $R_{min} = 0$ and $R_{max} = R_{fix}$
2. the probability of detecting the source for each acquisition time interval along the path

of the detector when the source is at distance R_{max} perpendicular to the path, as long as the probability for each time interval is larger than a chosen very small value, is calculated
3. all probabilities along the path to a total combined probability of detecting the source are combined. If the total combined probability P_{test} is higher than the desired probability $P_{desired}$, then the distance R_{max} will be doubled and returned to step 2
4. the lower threshold R_{min} and an upper threshold R_{max}, between which the maximum detection distance R_{detect} for the desired detection probability $P_{desired}$ is located. Then $R_{test} = (R_{max} - R_{min})/2$ is calculated, along with the total combined probability P_{test} of detecting the source at this distance
5. if $P_{test} - P_{desired} > 0.00001$, then $R_{min} = R_{test}$ is set, and step 4 is repeated. If $P_{desired} - P_{test} > 0.00001$ then $R_{max} = R_{test}$ is set, and step 4 is repeated
6. if the probability change $| P_{test} - P_{desired} |$ is less or equal to 0.00001, the calculation is halted and the MDD $R_{detect} = R_{test}$ is set

The process proceeds by refining the detection distance within the lower and upper thresholds, denoted as R_{min} and R_{max}, respectively. This refinement is achieved through a binary search approach aimed at narrowing down the maximum detection distance, R_{detect}, to achieve the desired detection probability, $P_{desired}$. This iterative refinement continues until the difference between the calculated and desired probabilities reaches a predefined threshold.

4. Results and Discussion

In this study, theoretical calculations were employed to determine the maximum detection distances achievable with a mobile NaI(Tl) spectrometer. Figure 2 illustrates the calculated distances corresponding to detection probabilities of 80%, 85%, 90% and 95%. The provided values pertain to the scenario where the source is positioned on the right side of the vehicle. It's worth noting that there exists a slight variance in the maximum detection distance contingent upon whether the source is situated on the left or right side of the vehicle.

The provided detection distances are based on a comprehensive analysis considering factors such as background radiation levels, source activity, detector efficiency, and angular response. Notably, these calculations focused on a Cs-137 point source, a commonly encountered radioactive material in environmental monitoring applications.

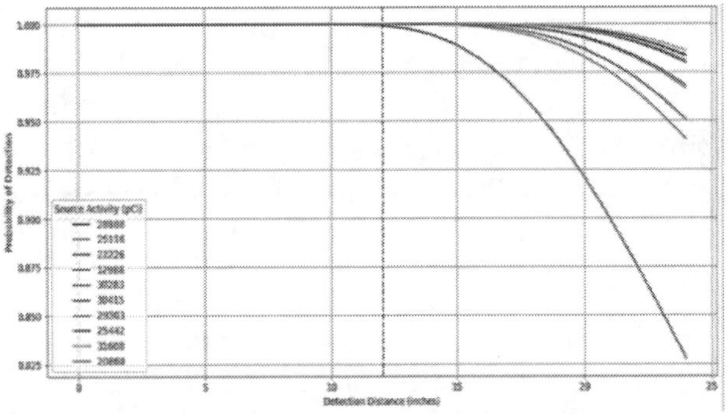

Figure 2. *The detectable distances with varying probabilities (80%, 85%, 90%, and 95%). These distances are governed by the MDD algorithm employed for a mobile detector.*

Choosing the appropriate acquisition time interval plays a pivotal role in maximizing the detection distances for point sources with a specific activity level. As depicted in Figure 3, the "photon intensity curve" exhibits varying widths along the trajectory of the measurement vehicle, contingent upon the distance of a radiation source from the roadway. A low-intensity source yields a photon signal measurable only over a short distance along the trajectory, with natural background radiation predominantly contributing to detector readings elsewhere. To detect the radiation source near the detection limit, the acquisition time interval must encompass a substantial portion of the width of the "photon intensity curve" under the least favourable alignment of acquisition time intervals.

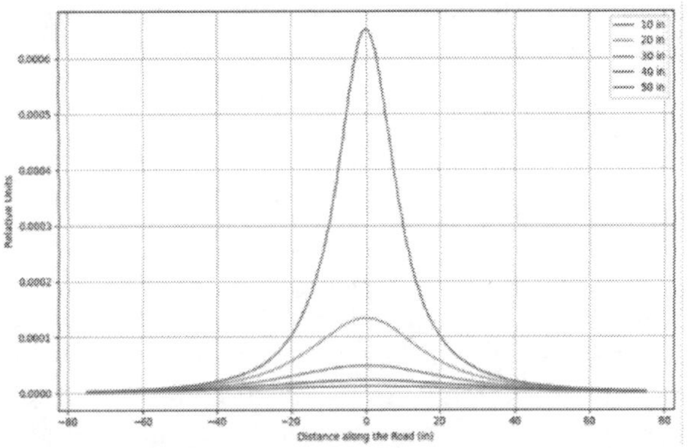

Figure 3: *The primary photon fluence rate along a linear path with a Cs-137 point source emitting photons at a*

primary energy of 662 keV positioned at varying distances (R = 10, 20, 30, 40, and 50 inches) from the path. These curves, governed by Equation 4, are exclusively influenced by the distance R separating the path from the source and the air attenuation coefficient pertinent to the primary photon energy.

The speed of the vehicle and the acquisition time intervals are intricately linked. At higher speeds, the detector traverses the peak along the "photon intensity curve" more rapidly, resulting in a reduction in the number of counts from the source in the detector and consequently shortening the maximum detection distance. In general, an increase in speed leads to shorter maximum detection distances when the acquisition time interval remains constant. However, simultaneous increases in both speed and the acquisition time interval further diminish the maximum detection distance.

Therefore, for sources with higher activity levels, employing a longer acquisition time interval to capture more photons from the source during each interval is more beneficial. The optimal acquisition time interval increases with rising source activity (and consequently distance) because a source situated farther away generates a broader "photon intensity curve" along the detector path.

The relative angular efficiency, δ_θ, according to Equation 8 was determined for the detector mounted in the measuring vehicle by obtaining the full energy peak count rates from the point source for $-180 < \theta < 180$. To account for the changing distances between the point source to the respective detector center, the count rates were corrected to achieve uniform intervals of 10 degrees. Figure 4 illustrates the angular efficiency characteristics observed in the detector.

While generally exhibiting uniformity across angles, the detector's placement within the instrument compartment leads to partial shielding, particularly noticeable for angles between $-20 < \theta < +20$ degrees. This decrease could indicate a shielding effect where the instrument or surrounding components obstruct the radiation path, leading to reduced detection efficiency in this angular range. Still, the detector maintains a somewhat uniform sensitivity to radiation for most incident angles within this range. Understanding the diverse angular efficiencies amongst different detectors is crucial for accurate measurement of photon fluence from different directions.

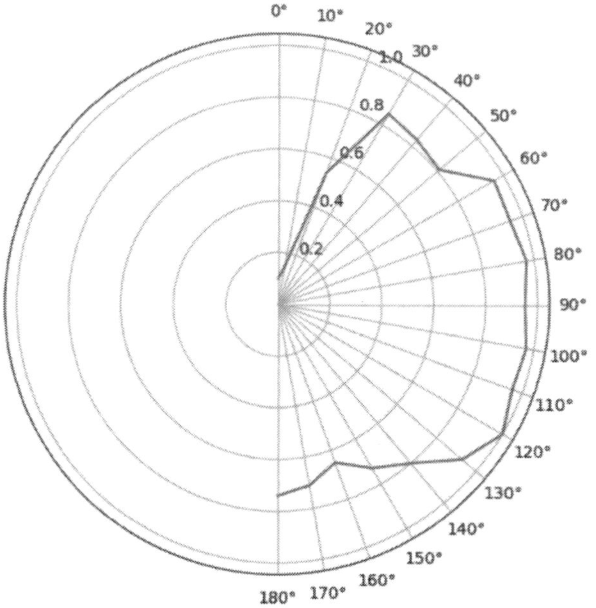

Figure 4. Relative angular efficiency, $\delta_\theta = N_\theta/N_{90}$, at 662 photon energy for 2" x 2" NaI(Tl) detector mounted on the right side of the vehicle with the audible volume facing backwards.

Variations in sensitivity across different detectors, or even within the same detector model may exist. These discrepancies could arise from manufacturing differences, calibration inconsistencies, or aging effects, potentially leading to inaccuracies in the measurement of photon fluence from different directions. Fluctuations in environmental parameters, such as humidity, temperature, and atmospheric composition can also affect radiation detection efficiency, attenuation, and background noise levels, introducing uncertainties into the measurement process.

5. Conclusion and Future Work

The maximum detectable distance signifies the farthest distance achievable between the vehicle and the radiation source under specific source characteristics and activity levels. This knowledge plays a pivotal role in assessing the detectability of the source within a designated radius. Implementing the MDD findings allows for the optimization of detector settings, vehicle velocities, and spectroscopy instrument configurations, ensuring more effective and efficient source-searching missions.

By identifying the most favorable acquisition time interval for any given combination of vehicle speed and source activity, the MDD algorithm aids emergency search teams in selecting the optimal parameter setup tailored to the specifics of each remedial undertaking. These insights are invaluable for operational planning and route design, facilitating the optimization of detection system deployment.

The determination of maximum detection distances in mobile gamma-ray spectrometry for post-fallout cleaning is contingent on various factors, including the specific radionuclide, the type and efficiency of the detector, vehicle speed, and the selected acquisition time interval. It has been found that optimal acquisition time intervals fall within the range of 5–10 seconds, particularly for achieving maximum detection distances in the case of Cs-137 when utilizing a 2" x 2" NaI(Tl) spectrometer. Increasing the speed of the vehicle has shown to reduce the maximum detection distances by approximately almost 15%. Consequently, a higher vehicle speed necessitates shorter acquisition time intervals to uphold the optimal conditions for achieving maximum detection distances.

Despite the utility of the MDD algorithm, it is essential to recognize the inherent limitations and uncertainties associated with its calculations. Assumptions regarding environmental conditions, source characteristics, and the accuracy of the theoretical model utilized may introduce biases necessitating further validation.

Future research endeavors could focus on refining the model, validating it through experimental data, and exploring additional factors that might influence detection distances – such efforts would enhance the reliability and applicability of our findings. Specifically, integrating terrain elevation data into the algorithm presents a promising avenue for future investigation. Analyzing how variations in terrain elevation impact gamma radiation detection could provide valuable insights into the influence of topographical features on detection performance.

Different parameter settings that influence the maximum detection distances in mobile search are reflected by the calculation results from the MDD algorithm, presenting the most favourable acquisition time interval aiding emergency search teams in selecting the most optimal parameter settings.

6. Acknowledgements

Our greatest gratitude to the New Jersey Bureau of Environmental Radiation, as well as the support of the New Jersey Radioanalytical Laboratory staff, whom this work would not have been possible without.

7. References

[1]Beck, H.L., DeCampo, J., Gogolak, C.V., 1972. In Situ Ge(Li) and NaI(Tl) Gamma-Ray Spectrometry for the Measurement of Environmental Gamma Radiation. USAEC Health and Safety Laboratory. Report HASL-258.

[2]B. Bird *et al.*, "A Robot to Monitor Nuclear Facilities: Using Autonomous Radiation-Monitoring Assistance to Reduce Risk and Cost," in *IEEE Robotics & Automation Magazine*, vol. 26, no. 1, pp. 35-43, March 2019, doi: 10.1109/MRA.2018.2879755

[3]Bukartas, A., Wallin, J., Finck, R., & Rääf, C. (2021). Bayesian algorithm to estimate position and activity of an orphan gamma source utilizing multiple detectors in a mobile gamma spectrometry system. PloS one, 16(1), e0245440. https://doi.org/10.1371/journal.pone.0245440

[4]Finck, Robert, et al. "Maximum detection distances for gamma emitting point sources in Mobile Gamma Spectrometry." *Applied Radiation and Isotopes*, vol. 184, 2022, p. 110195, https://doi.org/10.1016/j.apradiso.2022.110195

[5]F. R. S. C. Hjerpe T, "Statistical data evaluation in mobile gamma spectrometry: an optimization of on-line search strategies in the scenario of lost point sources.," Health Physics, vol. 80, no. 6, pp. 563-570, 2001.

[6]Gabrlik, Petr, and Tomas Lazna. "Simulation of gamma radiation mapping using an unmanned aerial system." IFAC-PapersOnLine, vol. 51, no. 6, 2018, pp. 256–262, https://doi.org/10.1016/j.ifacol.2018.07.163.

[7]Gunatilaka, A.H., Ristic, B., Skvortsov, A.T. and & Morelande, M.R., "Parameter estimation of a continuous chemical plume source," in 11th International Conference of Information Fusion, 2008.

[8]Howse, J.W., Ticknor, L.O. and & Muske, K.R., "Least squares estimation techniques for

position tracking of radioactive sources," Autom, vol. 37, pp. 1727-1737, 2021.

[9] Knoll, G. Radiation Detection and Measurement, 4 ed., John Wiley & Sons, Inc., 2010.

[10] Kock, P., Finck, R. R., Nilsson, J. M., Ostlund, K., & Samuelsson, C. (2010). A deviation display method for visualising data in mobile gamma-ray spectrometry. Applied radiation and isotopes: including data, instrumentation and methods for use in agriculture, industry and medicine, 68(9), 1832–1838. https://doi.org/10.1016/j.apradiso.2010.04.004

[11] Morelande, M.R., Ristic, B., & Gunatilaka, A.H., "Detection and parameter estimation of multiple radioactive sources," in 10th International Conference of Information Fusion, 2007.

[12] Peterson, John, et al. "Experiments in unmanned aerial vehicle/Unmanned Ground Vehicle Radiation Search." Journal of Field Robotics, vol. 36, no. 4, 12 Mar. 2019, pp. 818–845, https://doi.org/10.1002/rob.21867.

[13] Pradeep Kumar, K.A., et al. "Advances in detection algorithms for Radiation Monitoring." Journal of Environmental Radioactivity, vol. 217, June 2020, p. 106216, https://doi.org/10.1016/j.jenvrad.2020.106216.

[14] Rao, N.S.V., Xu, X. and & Sahni, S., "A Computational Geometry Method for DTOA Triangulation," 2008. [Online]. Available: https://apps.dtic.mil/sti/pdfs/ADA522040.pdf. [Accessed 20 08 2023].

[15] Sanada, Yukihisa, and Tatsuo Torii. "Aerial radiation monitoring around the Fukushima Dai-Ichi nuclear power plant using an unmanned helicopter." Journal of Environmental Radioactivity, vol. 139, Jan. 2015, pp. 294–299, https://doi.org/10.1016/j.jenvrad.2014.06.027.

[16] Song, Hankyeol, et al. "A radioactive material monitoring system using multiple gamma spectroscopy detectors and centroid method." Applied Radiation and Isotopes, vol. 128, Oct. 2017, pp. 36–40, https://doi.org/10.1016/j.apradiso.2017.06.029.

[17] Xu, X., Rao, N.S., & Sahni, S., "A computational geometry method for localization using differences of distances," ACM Transactions on Sensor Networks, vol. 10, pp. 1-10, 2010.

53e Biochemical Kinetics-Informed Ann Modeling for Cyanobacterial Biological Processes

Miriam Nnadili, Fangyuan Ma, Wei Sun, Michael G. Benton, Jose A. Romagnoli

C-phycocyanin (C-PC), an algal bio-pigment, is in high demand in the food and pharmaceutical industries due to its anti-inflammatory, neuroprotective, and antioxidant properties. Currently, C-PC is primarily obtained through extraction and purification from cyanobacterial cultures. However, the limited biomass concentration and phycocyanin productivity in photobioreactors (PBRs) have hindered the commercialization of C-PC production. To enhance C-PC yield, fed-batch optimization techniques are employed to control cyanobacterial growth. Typically, these optimization studies rely on first-principles models that accurately predict the behavior of the biosystem. Yet, due to the intricate metabolic mechanisms in microorganisms, creating precise first-principles models based on metabolic kinetics can be exceptionally challenging. To address this challenge, artificial neural networks (ANNs) have recently been applied to simulate and control dynamic bioprocesses. ANNs offer an advantage by treating the bioprocess as a 'black box,' constructing models solely from data. However, this approach has its strengths and limitations. While ANNs can establish models for biological processes, their accuracy depends on the range of available modeling data. Since biochemical kinetics information is not incorporated into ANN modeling, the relationships among variables in ANN-based models may not always be accurate. There has been limited emphasis on improving interpretability and generalization in previous studies.

In this study, we propose an ANN modeling approach for cyanobacterial biological processes that considers biochemical kinetics. To infuse the neural network with biological process information, we convert biochemical kinetics into constraints and embed them into the loss function. By minimizing this new loss function, we optimize the neural network parameters, allowing the model to converge towards the true biological process state during training. We validate our proposed method using experimental data from existing literature.

1a Regeneration: Fuel for the Future

Mark Lashier

Many industries are in the middle of rapid, disruptive, once-in-a-lifetime change. In today's business world, we need to embrace that change to regenerate, renew and grow. In his keynote, Phillips 66 President and CEO Mark Lashier will talk about what that journey means for his company and the energy industry, and why it's essential to maintain a culture of safety through it all.

AUTHOR INDEX